Become that

Business Analyst

Land that job sooner!

Ochuko Aluya

ISBN-10: 151463001X
ISBN-13:978-1514630013

DEDICATION

This book is dedicated to all who seek to become a Business Analyst (BA) .

CONTENTS

Acknowledgments   i

1   Introduction   3

2   What is Business Analysis   Pg. 4

3   The Natural Skills that makes you fit to be a BA   Pg. 7

4   Realistic approach to becoming a BA   Pg. 8

5   Eight (8) MUST HAVE skillset of a BA with examples   Pg. 10

6   Now that you are hired – Duties of a BA   Pg. 36

7   You are now a BA   Pg. 45

7   The Way Forward – Resume Construction   Pg. 46

8   Course Takeaway   Pg. 46

ACKNOWLEDGMENTS

Thank you to all who contributed to the success of this book.

1 INTRODUCTION

There are countless number of students who have graduated from college and are seeking earnestly to get insight on how to become a Business Analyst, but all they get is half baked knowledge. They are stranded somewhere in their career, not knowing what the job of a Business Analyst entails.

There are also a number of persons who are professionals who do not have complete knowledge of what a BA position entails. This course has come at the right time. Hopefully, students would be able to navigate life, functioning in the role they have always dreamed of – **Business Analyst**.

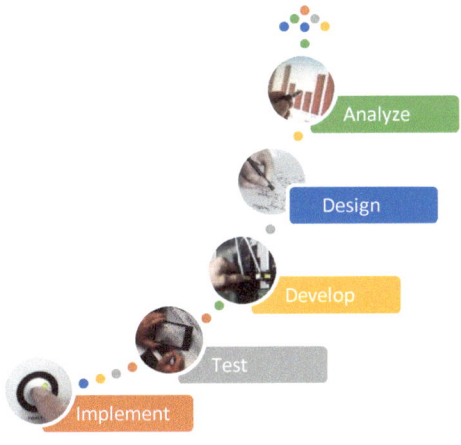

2 WHAT IS BUSINESS ANALYSIS?

Business Analysis is <u>not</u> the same as **System Analysis**. Persons who study Business Analysis are called Business Analyst, and persons who study System Analysis are called Systems Analyst.

In large organizations the role of a BA and SA overlap, but they are two different entities when it comes to skillset. The skillset of a *BA leans more towards the business side, while the SA leans towards the technical side*. In big firms, the BA role and SA role are clearly defined, but in smaller firms an individual performs the duties of a BA and SA.

Business Analysis is the process of identifying business needs and proffering solutions to them. Solutions include business process change, system redesign, re-engineering, etc.

System Analysis is the study of application/system procedure with the aim of diagnosing if it serves the business most appropriately, and defining the most efficient application schema to obtain a desired result for the business.

.

BA & SA Role

BA	SA
Identified as a field within business	**Identified as a field within IT**
• Emphasis is on **business knowledge**	• Emphasis on **system(s) knowledge**
• **Liaison** between the business team and IT professionals	• **Liaison** between IT and business professionals
• Requirements **elicitation**	• Requirements **implementation**
• Interface, **middle-person**	• Designer, developer, architect, **solutions-person**
• Skills in analysis, communication, **facilitation**	• Skills in analysis, design, **problem solving**
• Emphasis on articulating **business needs**	• Emphasis on supporting **application/business performance**

Business Analysis	Measure	System Analysis
Requirements Elicitation		Requirements Elicitation
Presentation		Presentation
Leadership		Leadership
Liaison/Documentation		Liaison
Business Knowledge		Business Knowledge
Communication		Communication
Creative Thinking		Creative Thinking
Problem Solving		Problem Solving
Technical Know-how		Technical Know-how

Level of Involvement in Projects

Large Scale

Small Scale

3 SKILLS THAT MAKES YOU FIT TO BE A BA

- o Problem Solving.
- o You have a strong passion to succeed big in an organization
- o You always want to know "Why"
- o You always seek excellence
- o Happy to clarify misunderstanding.
- o You like to document happenings
- o Love change
- o You are excited to learning new things
- o Proactive
- o You hate waste
- o You enjoy speaking
- o You enjoy working independently and in a team
- o You like to explore new things
- o You are detail oriented.
- o Patient
- o People person
- o You are seeking a new career
- o You are focused, Etc.

4 REALISTIC APPROACH TO BECOMING A BA

One Way: You could be functioning in the capacity of an application Quality Assurance (QA) personnel, which allows you to have solid understanding of the business processes and how the system performs. This is because you listen to the user, and you work with the system.

As a user's advocate (QA), you will naturally on the job start analyzing business processes and user's expectations of the system. This is an opportunity for you to start looking at internal job postings and find your way to the Business Analyst position you desire

Another Way: You may have discovered your natural skillset and realize just how it meets the skillset of a Business Analyst.

- Good listener, Patient, People person,
- Analytical, Problem solving, Enjoy speaking,
- Question why, Learn new things, etc.

All you need do, is sit for <u>PAID trainings no matter how expensive, study hard afterwards</u>. After, the training, contact Technical Recruiters to search for a Business Analysis job for you. Prepare for the Interview and, voila…the job is yours!

Yet Another Way: If you have been a **programmer** for many years, you would realize that you have been speaking to stakeholders on all levels as a result of the application defects logged from time to time. Your natural inclination is to the Front End. Since you are **aka "problem solver"**, you already understand the business process via the design of business applications.

If you require a change in career, all you need do study a bit, then apply for **internal job postings,** and get the job.

5 EIGHT (8) SKILLSET OF A BA – GETTING HIRED

- o Eliciting (get) Requirements
- o Documentation
  - o Business Requirement Document (BRD)
  - o Functional Requirement Document (FSD)
  - o Project Charter
  - o Problem Statement
  - o Vision and Scope
  - o Requirement Management Plan
  - o Retirement Traceability Matrix
- o Diagramming
  - o Business Data Modelling (BDM)
  - o Business Process Modelling (BPM)
  - o Mockup/Prototype
  - o Wireframe
  - o Sequence Diagram (SD)
  - o Use Case (UC)
  - o Activity Diagram
  - o Swim lane Diagram
  - o Entity Relationship Diagram (ERD)
- o Conducting JAD Sessions
- o Knowledge of Design Tools
- o Change Request Management
- o Development Methodology
  - o Joint Application Development (JAD)

- o Rapid Application Development (RAD)
- o Rational Unified Process (RUP)
- o Systems Development Life Cycle (SDLC)
- o Waterfall (a.k.a. Traditional)
- o AGILE/SCRUM
- o Knowledge of Testing

ONE - ELICITING (GET) REQUIREMENT

Requirement defines precisely what needs to be **created or changed.** This is one primary duty of a BA.

If a business wants to **create a new application or change certain features of an application** like the one on below.

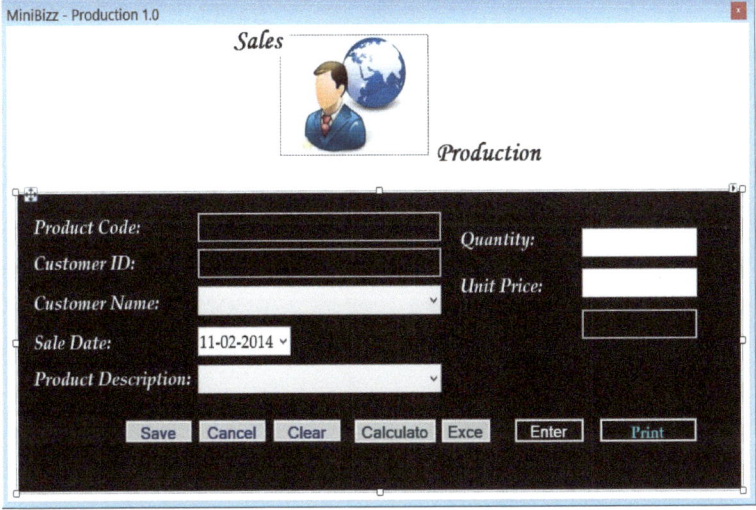

A BA needs to elicit requirements from the stakeholders (everyone that has a stake in the business) of the business, and document same.

Requirement documentations are written clearly with diagrams such as flowchart, process flow, mockups (User Interface), for the purpose of clarification. Elicitation can be done using various requirement elicitation technique such as;

- o SME's, Mgmt., etc)
    - o Document Analysis (reviewing existing documentation)

- o Interface Analysis (reviewing existing user interfaces)
- o Interviewing stakeholders
- o Observing (shadowing) stakeholders as they do their job
- o Prototyping
- o Joint Application Development (JAD) or Joint Requirement Plan(JRP) Sessions
- o Questionnaires

After elicitation of relevant requirements they have to be **committed to paper** for all to make reference to. There are several documents that a BA would have to create, one of them is the **Business Requirement Document (BRD)** used for **documenting business requirements.**

TWO - DOCUMENTATION

Documenting relevant information allows stakeholders to make reference to the documents if they are in doubt as to what the requirements are for the project. Also, documentation is necessary to keep projects in perspective, else people will deny facts, and argue endlessly about what was said and what was not said during important stakeholder deliberation. They are several important documents used by firms in some *Projects*... some are used to this day, but some are not used in certain projects. Regardless, it is very important for a BA to familiarize themselves with the relevant documents outlined below.

- o Business Requirements Document (BRD)
    - o Non Functional Specification
    - o Functional Specification
- o Functional Requirement Document (FSD)
- o Business Case (BC)
- o Project Charter (PC)
- o Problem Statement (PS)
- o Vision and Scope
- o Requirement Management Plan (RMP)
- o Requirement Traceability Matrix (RTM)

Business Requirements Document (BRD)
This document is written by a BA from the perspective of the business, outlining functional (what the system should be able to do) and non-functional (security, performance, usability, etc) requirement . **BRD**

specifies the need of a particular business application and the **focus is on what is required**, rather than on how to achieve it, which is usually relegated to a Systems Requirements Specification or Document (SRS or SRD) or Functional Specification Document.

Requirement documentation contains scope of work, how it will be accomplished, by whom, effort required, out of scope (what the project will not include), constraints, cost, budget, and other additional but relevant information which includes **Gap Analysis** (current versus future state of the application) depicted with flowchart, process flow, mockups (User Interface), for the purpose of clarification.

Get the link sample at this link - **http://1drv.ms/1QNciTW**

Functional Requirement Document (FSD)

This document is written by a BA from the perspective of the technical stakeholders (developers, programmers, designers etc.). It is a more detailed document that is **transcribed from the BRD**. It captures step by step **technical functionality** of the entire requirement outlined in the BRD. Contents include but not limited to;

- o Technical specification of all requirement list in the BRD using diagrams
    - o Use Case, Sequence, Process Flow, Data Flow
    - o Activity Diagram, Prototype/Wireframes
    - o Entity Relationship Diagram – Conceptual & Physical Elements, etc.

Get the link sample at this link - **http://1drv.ms/1e4866x**

Business Case Document

This a formal document written by a BA. It is a comprehensive document that outlines the justification for embarking on a project; the intent is to convince a decision maker to approve the project. It outlines feasible solutions to a given problem. This enables the mgmt. stakeholder to select the option that best serves the organization. It also, outlines time, human resource, and the financial impact/Return on Investment (ROI) the project will have on the business. Below are steps to take when writing a business case;

- o Do a thorough analysis (understand the problem)
- o Develop the "Executive summary"

- o Develop "The problem statement"
- o Develop "Analysis of the situation"
- o Develop "Solution options"
- o Develop "Project description"
- o Develop "Cost-benefit analysis"
- o Develop "Recommendations"

Get the link sample at this link - **http://1drv.ms/1KXlcwm**

Project Charter (PC)

A project charter (PC) is a document that is written by a Project Manager (PM). This document outlines the fact that a project exists, it presents to the project manager a written authority to begin work. It is important a BA knows what this document entails in the event the onus falls on them to prepare it.

It also assist the PM to communicate his expertise to project stakeholders (participants) why the project is embarked upon, who is involved, milestone, cost, what resources are required etc.

Content includes, but not limited to;

- o Project goal – reason the project is embarked upon.
- o Project participants and their roles.
- o Stakeholders – identifies mgmt. participant – need to know.
- o Requirements - identifies resources required for the project's to succeed.
- o Constraints - documents potential roadblocks.
- o Milestone - identifies timeline (start and end dates).
- o Communication - specifies project communication procedures.

Deliverables - specific products upon completion

Get the link sample at this link - **http://1drv.ms/1KXmRlG**

Problem Statement (PS)

The PM writes the problem statement. Sometimes, it is written as part of the Project Charter (PC). A problem statement is a clear and brief description of the issues that needs to be addressed by a by the project stakeholders. It is used to keep the team on track during the project effort. It is also used to validate that the effort delivered an outcome that solves the problem statement…a problem well stated is half solved. ~ Wallis Davis. Contents include but not limited to;

- o Original problem
- o Stakeholders who are most affected by the problem
- o Type of problem
- o Suspected cause of the problem
- o Goal for improvement and long-term impact
- o Proposal for addressing the problem
- o Final problem statement

The problem statement should take a realistic amount of time to articulate, design and deploy. A problem statement is developed using the 5W's;

- o **Who** - Who does the problem affect? Customers, etc.
- o **What** - What impact is the issue causing? - What would happen when it is fixed? What would happen if its not fixed?
- o **When** - When does the issue occur? - When does it need to be fixed?
- o **Where** - Where is the issue occurring? Processes, products, etc.
- o **Why** - Why is it important that the problem is fixed? What impact does it have on the business/customer?

Get the link sample at this link - **http://1drv.ms/1L8hK4u**

Vision & Scope (VS)

This document is written by a Project Manager. When a project begins, the Project Manager (PM) has an exclusive role to play. The PM is responsible defining the scope of the project at the start of the project. All stakeholders involved in the project have some input on scope, but the PM is solely responsible for its outcome. Vision and Scope document prepares the project for kick-off.

A well written vision and scope document will assist the project and its stakeholders avoid some of the problems projects are likely to encounter. A copy of the final document is given to every stakeholder involved with the project. The project manager ensures that everyone are on the same page with regards to the final product/software meeting the mgmt. stakeholders stipulated business requirement.

The document contains;

Business Requirements
- o Background
- o Business Opportunity
- o Business Objectives and Success Criteria

  o Customer/Market Needs, & Business Risks
Vision of the Solution
  o Vision Statement
  o Major Features, Assumptions and Dependencies

Scope Releases and Limitations
  o Scope of Initial Release, Subsequent Releases
  o Limitations and Exclusions

Get sample at this link - **http://1drv.ms/1FK7ZUf**

Requirements Management Plan (RMP)

This document is written by a BA. It identifies the process and procedures used to plan develop, monitor, and control requirements in all stages of a project's lifecycle. This document is the foundation on which all project requirement management policies and procedures are drawn.

Get sample at this link - **http://1drv.ms/1FK8sFU**

Requirements Traceability Matrix (RTM)

This document is written by a BA. It is a Quality Control document that focuses on one hundred (100%) coverage of business requirement. It helps in creating a snap shot to identify coverage gaps.

The testers will always draw up a test plan which contains Test ID, Test Scenario's and Test Cases. The job of the BA is to be able to use the Test ID and Test Scenario to map to each business requirement that the Test Scenarios are addressing using the business document requirement section number or Use Case ID. It will look something like the document in the next sheet.

Get sample at this link - **http://1drv.ms/1B3Jipe**

Get sample at this link - **http://1drv.ms/1B3JoNV**

Get sample at this link - **http://1drv.ms/1GxurCE**

See Sample below

Business Requirements

Business Requirement ID	Requirement Description	Comment
Customer Interface		
R01	All customers should be able to fill out and submit a query in a form-like page	Form on page
R02	Query form should have certain fields; name, gender, dropdown-query, complaint, text box/message, and phone # field.	Text box 300 char, name, req field name, dropdown,
R03	All customer should be able to select the type of query to submit – Query or Complain	Dropdown
R04	There should be a mechanism to submit option/query or complaint	
R05	Customer sees it is submitted successfully	Successful message sent
R06	Error messages displayed if query does not get submitted and message, "please resubmit"	

Sample Test Scenarios & Test Cases.

Use Case ID:	UC-OCSS001
Use Case Name:	Create Query

Created By:	Ochuko A.		**Last Updated By:**	Ochuko A.
Date Created:	2/4/15		**Last Revision Date:**	2/4/15

Actors:	Customer
Description:	Describes the process for creating query in a customer service system
Trigger:	Customer selects "Create Query" Menu Option
Preconditions:	1. Customer is already logged in
Post-conditions:	Query is successfully submitted
Normal Flow:	1. System presents list of menu options for customer 2. Customer selects Create Query 3. System presents an open query form 4. Customer fills out, name, gender, query reason, and message 5. Customer submits
Alternative Flows:	In step 5 of the normal flow, when the customer submits form 1.System display, "message successfully submitted" 2. Customer logs out 3.System terminates
Exceptions:	In step 5 of the normal flow, if messages submission fails 1. System will display a message "Query submission failed, please re-submit"
Business Rule:	Name & Dropdown field are required, Message field 300 char,
Includes:	None
Frequency of Use:	500 per day
Special Requirements:	None
Assumptions:	Everyone understands English
Notes and Issues:	None

Sequence Table for Flowchart

Series: 001	Business Area: Customer Interface	Owner: Operations Mgr
Process #: 001	Process Name: Create Query	
Sequence #	Step	Branch To
1	Customer creates query	
2	Customer Submits query	
3	Data Integrator receives query	
4	Data Integrator sends it to database	
5	Database receives data	
	If data successful, database sends successful msg to data integrator	
5	If message successful, data Integrator sends msg to customer	7
6	If message failed, data integrator sends error msg to customer, and ask to re-submit query	2
7	End	

Get sample Test Scenarios at this link - **http://1drv.ms/1B3Jipe**

Get sample RTM at this link - **http://1drv.ms/1B3JoNV**

THREE - DIAGRAMMING - UML

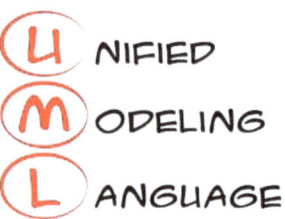

As a BA, you must be able know the concept of these design and to design all of these diagrams;

- o   Business Data Modelling (BDM)
- o   Business Process Modelling (BPM)
- o   Mockup/Prototype
- o   Wireframe
- o   Sequence Diagram (SD)
- o   Use Case (UC)
- o   Activity Diagram
- o   Swim lane Diagram
- o   Entity Relationship Diagram (ERD)

BDM also known as Data Flow Diagram (DFD)
Used for mapping the flow of data between external entities and the system.

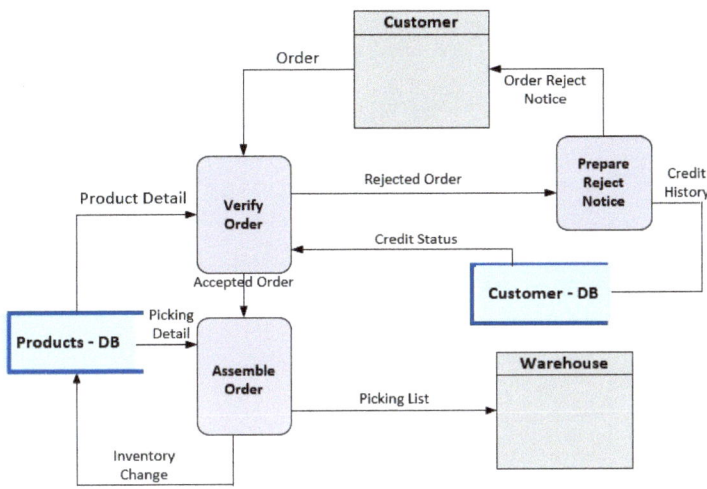

BPM also known as Process Flow Diagram

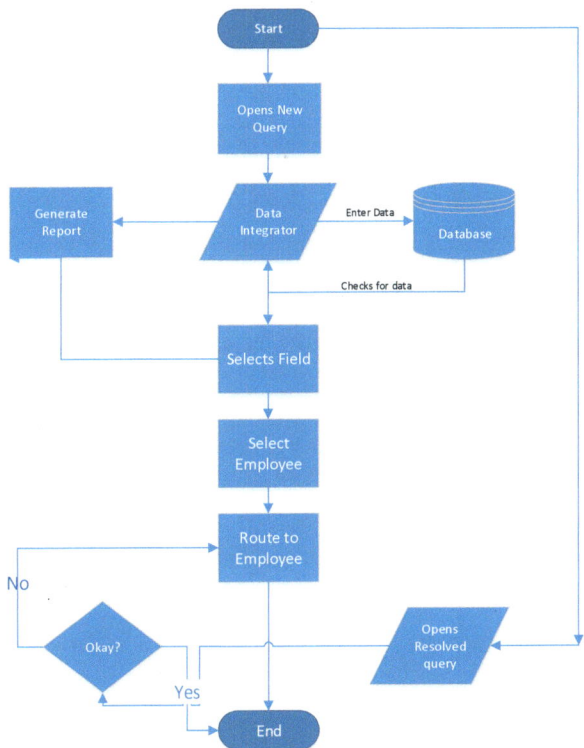

Business Requirement – Mockup/Prototype/Blue Print

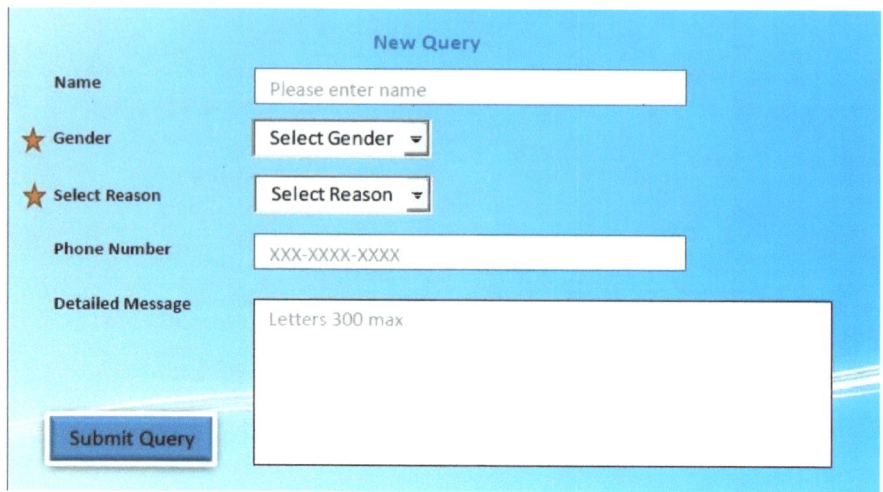

Business Requirement Wireframe – Skeletal Structure of the User Interface

Name

Gender

Select Reason

Phone Number

Detailed Message

SUBMIT

Sequence Diagram (SD)

Sequence Diagram - defines event sequences that result in some desired outcome.

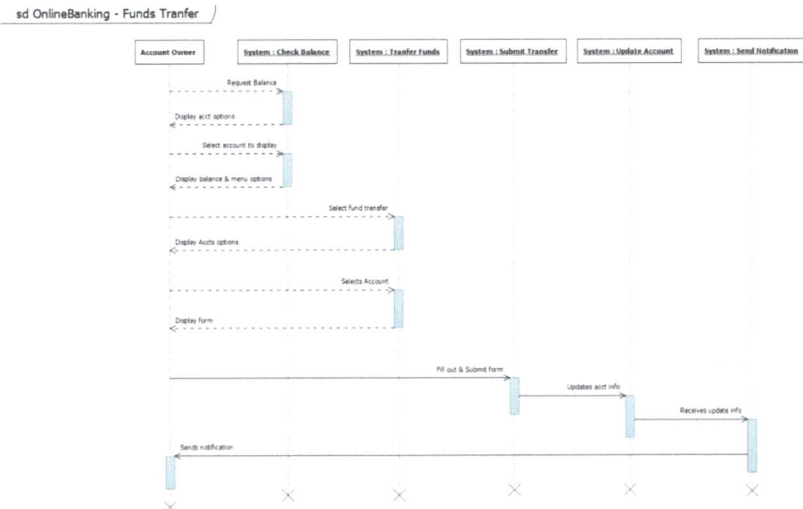

Use Case (UC) – Bank Funds

Each UC imply a sequence of steps involved in accomplishing the use case

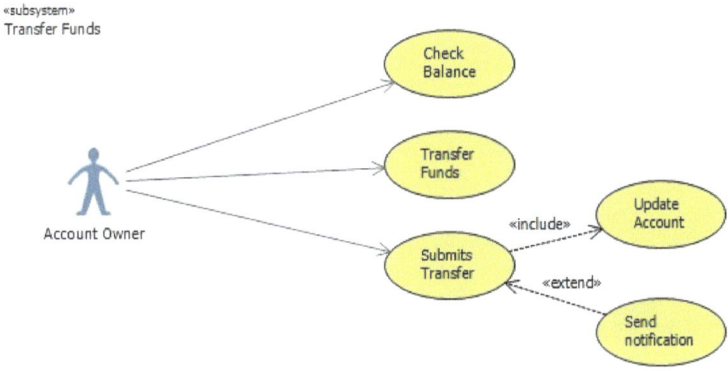

Activity Diagram

It is used to describe business processes that describe the functionality of the business system.

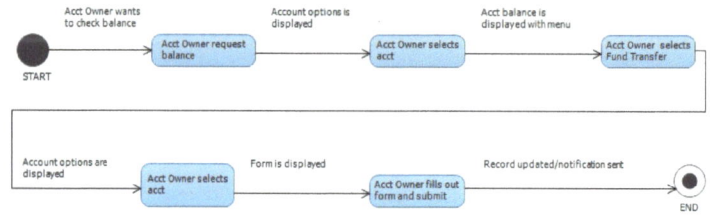

Entity Relationship Diagram (ERD) – Logical Diagram

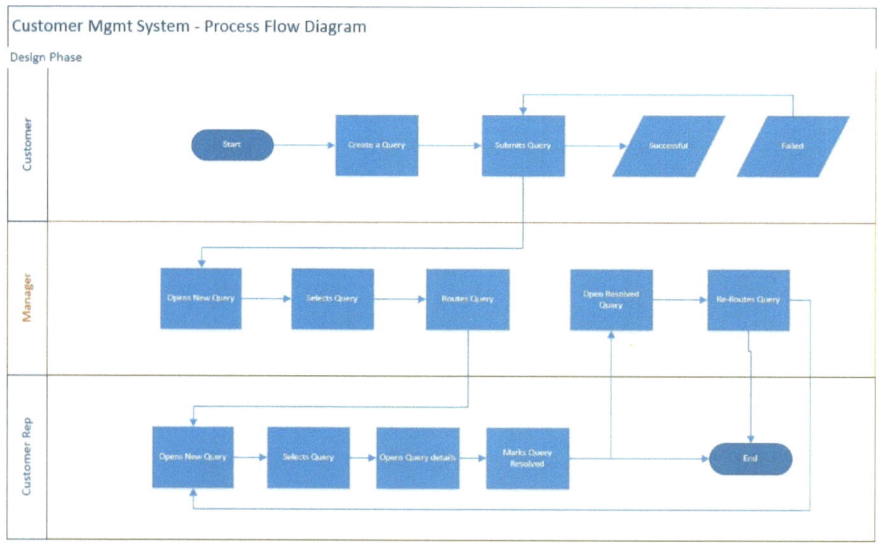

Entity Relationship Diagram (ERD) – Logical Diagram
Used for structuring a database schema

ENTITY RELATIONSHIP DIAGRAM

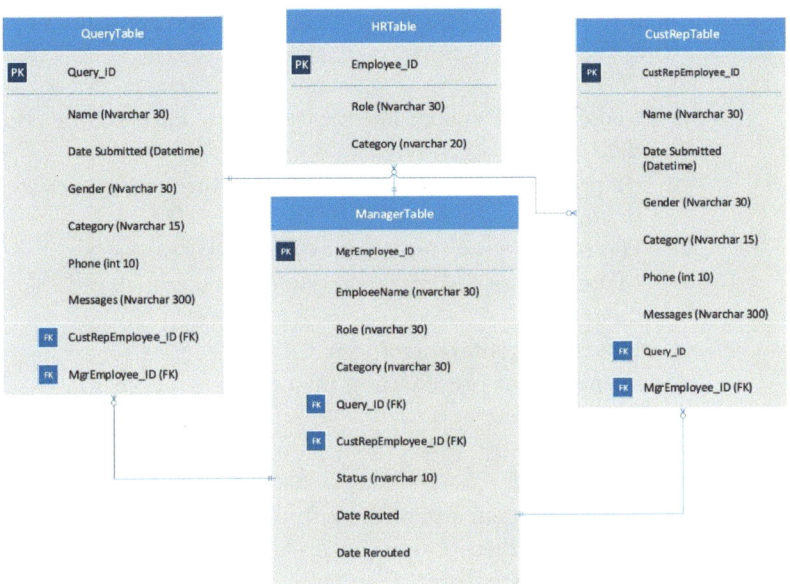

FOUR - CONDUCTING JAD SESSIONS/INTERVIEWS

JAD means Joint Application Development. It is the primary responsibility of a BA to organize this kind of gathering of key stakeholders to JOINTLY develop a solution and clarify business requirement which speeds up the application design process. Before conducting a JAD Session you should make sure you are ready using the following checklist;

- Make sure all stakeholders calendar are clear on the chosen date
- Be prepared – Design your PowerPoint presentation document
- Have your questions ready before the meeting
- Make a list of SME's/stakeholders that will attend the meeting.
- Reserve a conference room.
- Prepare your presentation materials e.g., projector, computer, etc.

Prepare your agenda;
- Send out email invitation with the agenda attached
- Prepare & distribution materials for review, then send via email 2days before.
- Send meeting reminder 1hr/1day before

On day of meeting;
- Assign a meeting clerk to take notes on your behalf
- Test your equipment and be sure it is working as planned 3hours before the meeting.
- Conduct the JAD Session

Example of questions you may ask;
- How do you perform your daily tasks with the system?
- How does this task affect the overall business functional areas?
- What are the different groups and roles involved in the function or process addressed by the product – how do they work together?
- Which other application interfaces with the one you use for your daily tasks?
- Can the design be implemented considering technology and environmental constraints?

- o Can the design be easy to migrate and integrate with another environment?
- o Are all critical functions identified?
- o Does the design address all issues from the requirements?
- o Does the design add features which was not specified in the BRD
- o Is requirement traceable to the design features?

Get sample Minutes of Minutes from this link - **http://1drv.ms/1GxuzSt**

FIVE - KNOWLEDGE OF DESIGN TOOLS
(At least know how to use one of each section)

- o **Requirement**
  - o Requisite Pro
  - o Requirement One etc.
  - o Axosoft – Product Backlog/Everything else
  - o MS TFS – Board/Everything Agile
  - o
- o **Use Case/Sequence diagram/Process flows/Task**
  - o UMLet
  - o MS Visual Studio Ultimate (Architecture menu)
  - o MS Visio etc.

Design tools (At least know how to use one of each section)

- o **Defects Tracking and Tasks**
  - o MS TFS
  - o Jira
  - o Rational Rose
  - o Rational ClearQuest
  - o Bugzilla
  - o Request Tracker
  - o Mantis
  - o Hotline etc.

- o **Test**
  - o HP ALM has a Requirement Traceability tool embedded

SIX - CHANGE REQUEST MANAGEMENT

If Stakeholders request a change using the stipulated form and submits it to the BA or sends an email. You will create a change request in Team Foundation Server or whatever tool the project has designated for that. Input the necessary info such as;

- o Change request description
- o Who requested the change
- o Change date
- o Change reason
- o Change request Status

Then;
- o Analyze it to figure out the business requirement and functional requirement
- o Pass it on to the development team to say if it can be done or not (determine feasibility) and how long
- o Forward to project manager to meet stakeholders and tell them more resources will be needed or not to get this done
- o PM approvals or decline, if not approved discard. If approved,
- o Once approved, it is added as a functional requirement to the FSD by the BA and work begins on it.
- o BA will categorize the requirement and decides if high priority or low priority
- o Status is updated from new to started, when complete it is changed to complete.

SEVEN - DEVELOPMENT METHODOLOGY

An Application or software **development methodology** is a roadmap used to structure, plan, and control the process of developing an application or software. There are several development methodologies but we will relegate our focus to the five topmost development methodology i.e.;

- o Joint Application Development (JAD)
- o Rapid Application Development (RAD)
- o Rational Unified Process (RUP)
- o Systems Development Life Cycle (SDLC)
- o Waterfall (a.k.a. Traditional)
- o AGILE/SCRUM

JAD (Please see topic on JAD Session)

JAD is a team based model. Its sessions can be a tedious process. Many organizations have adopted JAD sessions exclusive for requirement modelling. The outcome of a JAD session is, a requirement model or a more accurate statement of system requirements, a better understanding of common goals, and a stronger commitment felt by users and major stakeholder.

Get sample of JAD Invite in this link - **http://1drv.ms/1L8iGpz**

RAD – Development Phases

User Design and Construction repeats until construction completes.

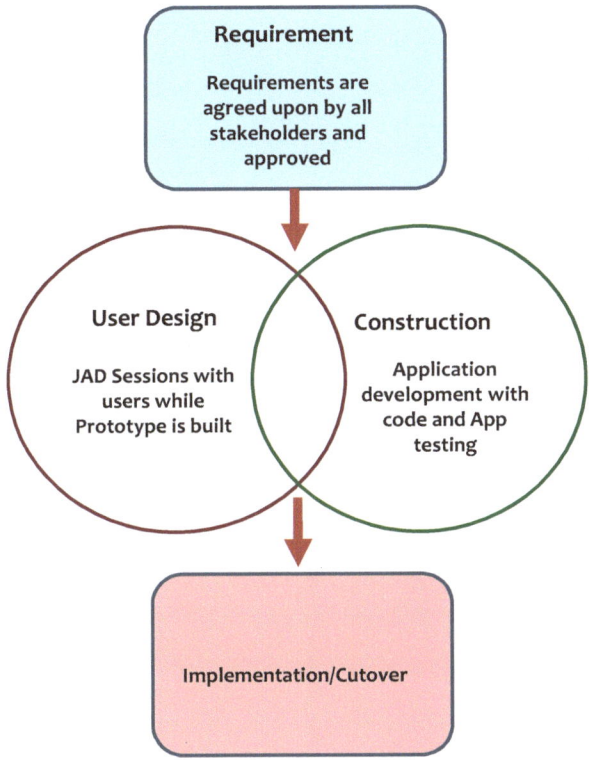

RUP

It is an **<u>Iterative</u>** software development process with four (4) Core Phases

<u>Iterative</u> means repetitive or to re-do: it is a development process for arriving at a desired result by **repeating rounds of each cycle**... It is a complete development loop resulting in a release of an executable product, a subset of the final product under development, which grows **incrementally from iteration to iteration** to become the final system. **The focus is on product quality.**

RUP – Development Phases

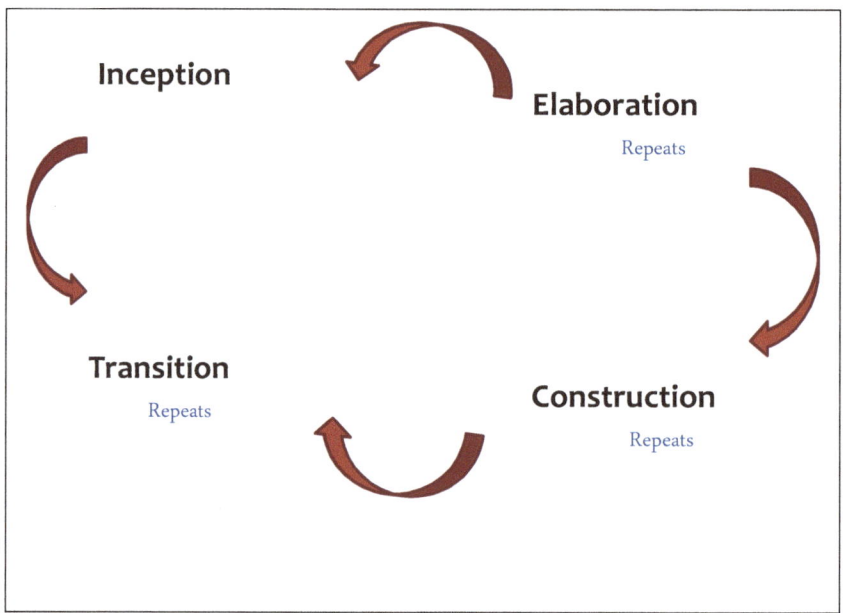

SDLC – Development Phases

System Planning

- o   Analyzing the Business Case

System Analysis

- o   Requirements Modelling
- o   Data Process Modelling
- o   Object Modelling

System Design

- o   User Interface
- o   Data Design
- o   System Architecture

Development

- o   Coding & Design

Systems Implementation

Waterfall

Waterfall is a popular systems development life cycle model. Its development method is rigid and linear. Each phase in the development cycle has specific deliverable, and one phase completes before the next begins. Schedule is set on stone, because of this, projects meet tight deadlines because each phase is carefully planned.

This model often fall short of expectations as it does not accommodate inevitable changes that often plaque most projects. Once the application is in the testing stage, it becomes difficult to go back and make a change when a defect is discovered.

Waterfall Development Phases

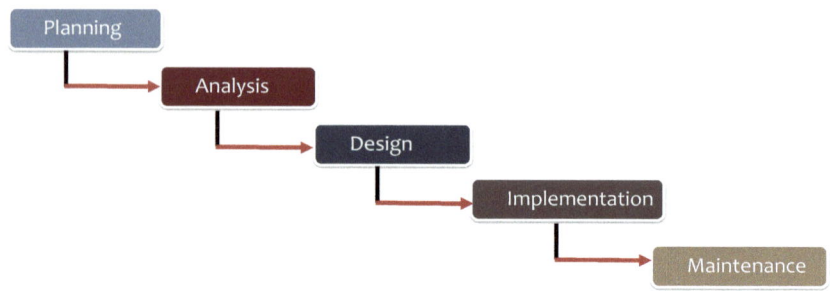

AGILE/SCRUM

Agile Methodology is an alternative for traditional software development methodology. The name Agile is used interchangeably with SCRUM. SCRUM or Agile is team-based, and the most popular form of developing software/application because of the simple fact that it is flexible and simplistic.

In Agile/Scrum there is no Project Manager (PM). The Product Owner, Team member and Scrum Master are the only three roles that exist in this model. The Scrum Master is responsible for facilitating the meeting, the Product Owner is responsible to providing the Backlogs (requirements), and the Team Members deliberate in the Backlog refinement (also known as backlog grooming) to come up with concise user stories (description of a feature in a proposed system), Sprint Planning, Daily Scrum, Sprint review and Sprint retrospective meetings.

AGILE/SCRUM – Development Phases
Process followed top down; daily scrum repeats daily

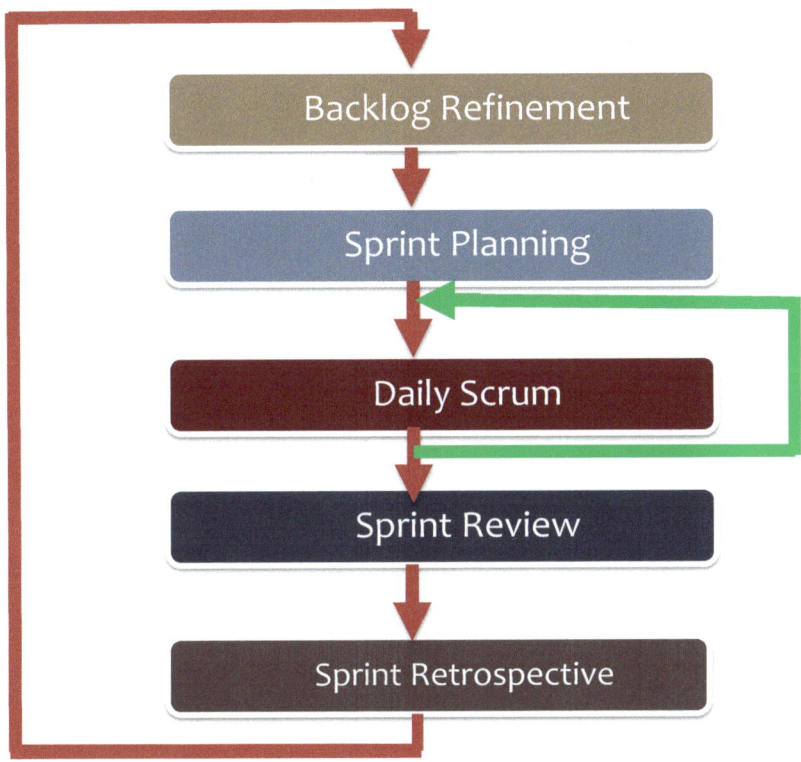

EIGHT - KNOWLEDGE OF TESTING

Testing is a procedure to validate the functionality of an application or system that it meets the client or business requirement.

The **BA's role** development of the application and testing effort make sure the business requirement are being met.

- o **Test Plan:** It is a document that describes the scope, method, schedule of testing activities, and classifies test items, testing tasks, who its assigned to and more.
- o **Test Step:** It is a written format for testing an application manually.
- o **Test Script or case:** is series of steps that testers follow to perform

manual testing on an application to verify its functionality.

- o **Test Scenario:** is a list of all possible areas to be tested in an application.
- o **Automated Testing:** This is testing carried out on a given application using test tools to automate the testing process.
- o **Manual Testing**: This is testing carried out on a given application without the use of test tools to automate the testing process. This is done using test case or script.
- o **Unit:** It is testing done on specific individual units of an application.
- o **Integration Testing:** is testing done on a complete application derived from individual units combined.
- o **Positive Testing:** is testing done on an application to determine if the system works.
- o **Performance Testing:** is testing done to measure the speed and efficiency of an application based on requirement stated.
- o **Usability Testing:** is testing to evaluate how the friendly the application front end interface is.
- o **Regression Testing:** is the testing done after a section of an application is modified, this is to ensure other related and underlying components not directly impacted by the modification still functions as it should.
- o **Blackbox Testing:** is testing validation technique that focuses on the output produced based on the inputs entered as opposed to the internal workings of an application.
- o **Whitebox Testing:** is testing verification technique that focuses on the internal workings of a system as opposed to output generated as a result of an input entered into an application; is also called structural testing or glass box testing.
- o **Stress Testing:** is testing done to calculate how an application functions under unusual conditions.
- o **Beta Testing:** is testing done by anyone outside of the development team on a the initial release of an application or product; the goal for error discovery.
- o **Negative Testing:** is test done to ascertain if an application is displaying unexpected error in a scenario of accurate application operation and vice versa.
- o **End-to-End Testing:** is testing done on the overall functionality of the system inclusive of data integration, to ensure data integrity within components of an application/system.
- o **Sanity Testing:** is testing done after modification to an application to ascertain that no further issues are introduced due to these changes.
- o **Functional Testing:** is testing done to ascertain that the system

functions the way it should.

- o **System Testing:** is the testing done to ensure an application works functions the way it should when put into production or operational environment.
- o **Non-functionality Testing:** is testing done to ascertain that the non-functional areas of the application such as security, performance etc. are working as expected.
- o **Smoke Testing:** is testing done on an application to ascertain the most important functional units are working as they should.
- o **User Acceptance Testing (UAT):** is testing done to ensure that the delivered product meets the client or business requirements.

NOW THAT YOU ARE HIRED – DUTIES OF A BA

Your initial mindset and attitude

- ○ Dress formal, Smile, be observant, and pay attention
- ○ Carry your analytical and interpersonal skill with you

Introduction To The Project Team

On your first day, you will be introduced by the pm to the **project team members.**

Stakeholders Identification

You may need to ask the PM some necessary questions to be in synch with the team sync

- o What is the project about – Overview? (re-engineering or a brand-new project.
- O What kind of development methodology in use?
- O What phase of development are they?
- O How many people make up the project team members?
- O Who are the developers, designers, dba's etc.?
- O Are there other ba's besides me?
- O Who are the key stakeholders, sponsors, sme's etc.?
- O Are there existing documents – brd, fsd etc.? (go through the documents)
- O Are there existing project templates or its build from scratch?
- O What project tools are in use?
- O Are there existing processes and procedures for communication? And more…

Analysis of stakeholder process – project matrix
This is the time to familiarize yourself with the project activity, and align yourself with the project – most especially get used to the project activity matrix – who is doing what.

- O   Project sponsor
- O   Project manager
- O   Business development managers
- O   Customers
- O   Project team members
- O   Support staff
- O   End user: who are the end users for uat
- O   Investors
- O   Brokers

Understanding the business processes
This is the time to know the **nitty-gritty of the business**. get details on what drives the business - **goal and objec**tives… we all know that every business exist for the purpose of satisfying customers and making profit. before an application for a client is built, a thorough analysis of the business will have to take place for the purpose of understand the model in place.

If you don't have an understanding of how the business works, you will not be able to **gather requirements accurately**. gathering requirements entails knowing the right questions to asks for business model purposes.

Business Modelling (BM) is the **analysis** and documentation of the **enterprise structure** - which is **the plan** implemented by a company to generate revenue and make a profit from operations. the model includes the **components/workings** and **functions/tasks** of the business, as well as **revenue and expense**.

Gap Analysis

Gap analysis (GA) is most useful at the beginning of a project. It compares the current state of the business model/application processes with the future state of the same. GA is employed to bridge the gap to produce a successful project in the end.

To carry this out, some steps will need to carried out with the objective and goal of the business in mind.

As-Is And To-Be – Done by BA's
(Solution through reengineering/restructuring)

Identify your future state: what is the objective that will take you to the complete application?

Analyze your current state: for each of the stated objectives, do an analysis of the current state of the application.

To do this, consider the following questions:
- O Who are the subject matter experts that can respond to question on the current state of the application?
- O How do i get the information from the experts; one-on-one interviews? Shadowing?

Bridge the gap: immediately both states are established by way of process diagram, work flows, data flows, and activity diagram, and tasks can now be organized in such a way that the projects objectives and milestone can be achieved.

Problem Statement

This document will be formally written based on the gap analysis derived, describing exactly what the problem is, and the proposal for address them by adopting the below lines;
- O What the original problem is
- O Stakeholders who are most affected by the problem
- O Type of problem
- O Suspected cause of the problem
- O Goal for improvement and long-term impact
- O Proposal for addressing the problem
- O Final problem statement.

Feasibility Study

Now, the pm will do feasibility study (process of analyzing the state of a given problem to get to a conclusion of whether or not a project can be undertaking, considering the cost, budget, time constraints, scope and many other elements) based on the problem statement.

Business Case

The BA writes the business case which outlines compelling reasons why the project should be undertaken, and if not adopted, compelling reasons why the business may fail;

- O Business problem
- O Problem analysis
- O Available options
    - O Recommended options
    - O Option rankings
    - O Option recommended
- O Implementation approach
    - O Project initiation
    - O Project planning
- O Project execution

Project Charter

After the business case is approved. The project manager writes the project charter (pc) which outlines the fact that a project exists, the project sponsors endorse it, and **this becomes a written permission for the project manager to begin work.**

Vision And Scope

This is a document written at the initial stage of the project, **the pm defines the scope of the project** through this document.
All stakeholders contribute to the scope. This will help the team to avoid some problems projects are likely to encounter.

You will be handed a copy of this document as everyone is handed a copy of it. This ensures that everyone is on the same page.

Requirement Management Plan (RMP)

The BA writes the RMP which identifies the **process and procedures used to plan develop, monitor, and control requirements at all stages of a project's lifecycle** following the lines below;

- O Introduction
  - O Purpose of the requirements management plan
- O Requirements management overview
  - O Organization, responsibilities, and interface.
  - O Tools, environment, and infrastructure

- O Requirements management
  - O Assumptions/constraints
  - O Requirements definition
  - O Requirements traceability
  - O Workflows and activities
  - O Change management
- O Requirements management plan approval

Requirement Elicitation

It is the primary responsibility of the BA to elicit requirement from the business stakeholders. Getting accurate requirement impacts the overall outcome of the final project. The goal is to identify all features required to develop the application. Requirement elicitation take various form, the BA has to choose the best technique that is suitable to getting the right requirements which will in turn be used to formulate the business requirement document (BRD);

- O Interviews, focus group, JAD session, questionnaire
- O Observation/shadowing, prototypes and more

You will write **JAD invite, and minutes of meetings with agenda** for your requirement sessions.

Writing The BRD Document

When all requirement are completely elicited from the business stakeholder, a formal documentation of the requirement will be done which is known as business requirement document (*see video on BRD and sample document*). BRD should contain all diagrams that will help in explaining complex requirement. The document will have to be reviewed and agreed upon if there are no specific changes to be made.

When the business stakeholder approves it, it becomes the scope of work (functional and nonfunctional requirement) written from the perspective of the business, but the content of this same document will have to be translated to an FSD which the developers/designers will work on.

Writing The FSD Document

The FSD document will now be drawn from the content of the BRD. The FSD will be written from the perspective of the technical stakeholders (developers, programmers, designers etc.) With more technical details specifying all requirement listed in the BRD using diagrams such as listed below and more;

- O   Use case, sequence, process flow, data flow
- O   Activity diagram, prototype/wireframes
- O   Entity relationship diagram – conceptual & physical elements, etc.

Requirement Traceability Matrix – Construction

As the developers begin design and development using the FSD, it is the responsibility of the BA to make sure requirement are not missed. This document known as quality control. It is like a GPS or compass that shows the way to a destination. It helps in creating a snap shot to identify coverage gaps.

When the testers draw up their test plan which contains, test scenario's and test cases. The job of the BA is to be able to use the test id's listed and test scenario to map to each business requirement that the test scenarios are addressing using the business document requirement section number or use case id.

Application Development

The developers will get to work developing/designing the application according to specification in the FSD. The developers design will be monitored by the BA to ensure they are strictly adhering to the stipulated requirement. The BA monitors development because, they are the advocate of the business stakeholders, and in turn the pm ensures everyone is following the timeline/schedule for the project. Once development is complete, testing will begins.

All Manner Of Testing Done

All manner of testing will begin in the development (DEV) and quality assurance (QA) environment to ensure the system functions as stipulated in the BRD. It is the responsibility of the BA to follow through with making sure this is done.

System Implementation

When the testing done in the development (DEV) and quality assurance (QA) environment meets requirement, the application is implemented in the UAT environment where selected users will test the system using test scripts written by the testers to make sure the system functions exactly as it should.

User Acceptance Testing – UAT

End users are selected to test the system using test scripts written by the testers to make sure the system functions exactly as it should. The test script are mapped to test scenarios and test cases. When a defect is found, a remark of "failed" will be written beside the test case that failed, and vice versa.

In all of this, the BA's job is to ensure the system/application is functioning according to the requirement in the BRD from the business stakeholders.

You are now a Business Analyst (BA)

- O   You are now equipped with the **tools necessary** to do your job as a **business analyst** – <u>take a step further to learn how to design those diagrams</u> using the tools shown to you in the previous course.
- O   You should now prepare very well for an interview by doing thorough research, practicing likely scenario based questions, and mastery of all business analysis terminology – then go get that **entry level job!**

Course take-away

- O Business and system analysis defined
- O Realistic way to get a job as a BA
- O Eight skillset of a BA

The Way Forward – Resume Construction

Make sure to construct your resume to reflect these skills only if you have mastered what the job of a BA entails, construct those diagrams, and practice likely interview scenario based questions. There are lots of YouTube videos that teaches you how to do that.

ABOUT THE AUTHOR

BS Computer Science with Statistics
MS Management Information Systems
Experience as a Business System Analyst.

www.ingramcontent.com/pod-product-compliance
Lightning Source LLC
Chambersburg PA
CBHW040917180526
45159CB00002BA/512